Y0-EIX-334

SPSS Regression Models™ 12.0

For more information about SPSS® software products, please visit our Web site at *http://www.spss.com* or contact

SPSS Inc.
233 South Wacker Drive, 11th Floor
Chicago, IL 60606-6412
Tel: (312) 651-3000
Fax: (312) 651-3668

SPSS is a registered trademark and the other product names are the trademarks of SPSS Inc. for its proprietary computer software. No material describing such software may be produced or distributed without the written permission of the owners of the trademark and license rights in the software and the copyrights in the published materials.

The SOFTWARE and documentation are provided with RESTRICTED RIGHTS. Use, duplication, or disclosure by the Government is subject to restrictions as set forth in subdivision (c) (1) (ii) of The Rights in Technical Data and Computer Software clause at 52.227-7013. Contractor/manufacturer is SPSS Inc., 233 South Wacker Drive, 11th Floor, Chicago, IL 60606-6412.

General notice: Other product names mentioned herein are used for identification purposes only and may be trademarks of their respective companies.

TableLook is a trademark of SPSS Inc.
Windows is a registered trademark of Microsoft Corporation.
DataDirect, DataDirect Connect, INTERSOLV, and SequeLink are registered trademarks of DataDirect Technologies.
Portions of this product were created using LEADTOOLS © 1991-2000, LEAD Technologies, Inc. ALL RIGHTS RESERVED.
LEAD, LEADTOOLS, and LEADVIEW are registered trademarks of LEAD Technologies, Inc.
Portions of this product were based on the work of the FreeType Team (*http://www.freetype.org*).

SPSS Regression Models™ 12.0
Copyright © 2003 by SPSS Inc.
All rights reserved.
Printed in the United States of America.

No part of this publication may be reproduced, stored in a retrieval system, or transmitted, in any form or by any means, electronic, mechanical, photocopying, recording, or otherwise, without the prior written permission of the publisher.

1 2 3 4 5 6 7 8 9 0   06 05 04 03

ISBN 0-13-109674-5

Preface

SPSS 12.0 is a powerful software package for microcomputer data management and analysis. The Regression Models option is an add-on enhancement that provides additional statistical analysis techniques. The procedures in Regression Models must be used with the SPSS 12.0 Base and are completely integrated into that system.

The Regression Models option includes procedures for:

- Logistic regression
- Multinomial logistic regression
- Nonlinear regression
- Weighted least-squares regression
- Two-stage least-squares regression

Installation

To install Regression Models, follow the instructions for adding and removing features in the installation instructions supplied with the SPSS Base. (To start, double-click on the SPSS Setup icon.)

Compatibility

SPSS is designed to run on many computer systems. See the materials that came with your system for specific information on minimum and recommended requirements.

Serial Numbers

Your serial number is your identification number with SPSS Inc. You will need this serial number when you contact SPSS Inc. for information regarding support, payment, or an upgraded system. The serial number was provided with your Base system.

Customer Service

If you have any questions concerning your shipment or account, contact your local office, listed on the SPSS Web site at *http://www.spss.com/worldwide*. Please have your serial number ready for identification.

Training Seminars

SPSS Inc. provides both public and onsite training seminars. All seminars feature hands-on workshops. Seminars will be offered in major cities on a regular basis. For more information on these seminars, contact your local office, listed on the SPSS Web site at *http://www.spss.com/worldwide*.

Technical Support

The services of SPSS Technical Support are available to registered customers. Customers may contact Technical Support for assistance in using SPSS products or for installation help for one of the supported hardware environments. To reach Technical Support, see the SPSS Web site at *http://www.spss.com*, or contact your local office, listed on the SPSS Web site at *http://www.spss.com/worldwide*. Be prepared to identify yourself, your organization, and the serial number of your system.

Additional Publications

Additional copies of SPSS product manuals may be purchased directly from SPSS Inc. Visit our Web site at *http://www.spss.com/estore*, or contact your local SPSS office, listed on the SPSS Web site at *http://www.spss.com/worldwide*. For telephone orders in the United States and Canada, call SPSS Inc. at 800-543-2185. For telephone orders outside of North America, contact your local office, listed on the SPSS Web site.

The *SPSS Statistical Procedures Companion*, by Marija Norusis, is being prepared for publication by Prentice Hall. It contains overviews of the procedures in the SPSS Base, plus Logistic Regression, General Linear Models, and Linear Mixed Models. Further information will be available on the SPSS Web site at *http://www.spss.com* (click Store, select your country, and click Books).

Tell Us Your Thoughts

Your comments are important. Please let us know about your experiences with SPSS products. We especially like to hear about new and interesting applications using the SPSS system. Please send e-mail to *suggest@spss.com* or write to SPSS Inc., Attn.: Director of Product Planning, 233 South Wacker Drive, 11th Floor, Chicago, IL 60606-6412.

About This Manual

This manual documents the graphical user interface for the procedures included in the Regression Models module. Illustrations of dialog boxes are taken from SPSS for Windows. Dialog boxes in other operating systems are similar. The Regression Models command syntax is included in the *SPSS 12.0 Command Syntax Reference*, available on the product CD-ROM.

Contacting SPSS

If you would like to be on our mailing list, contact one of our offices, listed on our Web site at *http://www.spss.com/worldwide*.

Contents

1 Choosing a Procedure for Binary Logistic Regression Models   1

2 Logistic Regression   3

Logistic Regression Data Considerations 4
Obtaining a Logistic Regression Analysis 4
Logistic Regression Variable Selection Methods. 6
Logistic Regression Define Categorical Variables 7
Logistic Regression Save New Variables 9
Logistic Regression Options 10
LOGISTIC REGRESSION Command Additional Features. 11

3 Multinomial Logistic Regression   13

Multinomial Logistic Regression Data Considerations. 13
Obtaining a Multinomial Logistic Regression. 14
Multinomial Logistic Regression Models. 15
Multinomial Logistic Regression Reference Category 17
Multinomial Logistic Regression Statistics 18
Multinomial Logistic Regression Criteria. 19
Multinomial Logistic Regression Options. 21
Multinomial Logistic Regression Save. 22
NOMREG Command Additional Features. 23

4 Probit Analysis 25

Probit Analysis Data Considerations 25
Obtaining a Probit Analysis 26
Probit Analysis Define Range 28
Probit Analysis Options 28
PROBIT Command Additional Features 29

5 Nonlinear Regression 31

Nonlinear Regression Data Considerations 32
Obtaining a Nonlinear Regression Analysis 32
Nonlinear Regression Parameters 34
Nonlinear Regression Common Models 35
Nonlinear Regression Loss Function 36
Nonlinear Regression Parameter Constraints 37
Nonlinear Regression Save New Variables 38
Nonlinear Regression Options 39
Interpreting Nonlinear Regression Results 40
NLR Command Additional Features 40

6 Weight Estimation 43

Weight Estimation Data Considerations 43
Obtaining a Weight Estimation Analysis 44
Weight Estimation Options 46
WLS Command Additional Features 46

7 Two-Stage Least-Squares Regression 47

Two-Stage Least-Squares Regression Data Considerations 48
Obtaining a Two-Stage Least-Squares Regression Analysis 48
Two-Stage Least-Squares Regression Options 50
2SLS Command Additional Features 50

Appendix

A Categorical Variable Coding Schemes 51

Deviation 51
Simple 52
Helmert 53
Difference 54
Polynomial 54
Repeated 55
Special 56
Indicator 57

Index 59

Chapter

1

Choosing a Procedure for Binary Logistic Regression Models

Binary logistic regression models can be fitted using either the Logistic Regression procedure or the Multinomial Logistic Regression procedure. Each procedure has options not available in the other. An important theoretical distinction is that the Logistic Regression procedure produces all predictions, residuals, influence statistics, and goodness-of-fit tests using data at the individual case level, regardless of how the data are entered and whether or not the number of covariate patterns is smaller than the total number of cases, while the Multinomial Logistic Regression procedure internally aggregates cases to form subpopulations with identical covariate patterns for the predictors, producing predictions, residuals, and goodness-of-fit tests based on these subpopulations. If all predictors are categorical or any continuous predictors take on only a limited number of values—so that there are several cases at each distinct covariate pattern—the subpopulation approach can produce valid goodness-of-fit tests and informative residuals, while the individual case level approach cannot.

Logistic Regression provides the following unique features:

- Hosmer-Lemeshow test of goodness of fit for the model
- Stepwise analyses
- Contrasts to define model parameterization
- Alternative cut points for classification
- Classification plots
- Model fitted on one set of cases to a held-out set of cases
- Saves predictions, residuals, and influence statistics

Multinomial Logistic Regression provides the following unique features:
- Pearson and deviance chi-square tests for goodness of fit of the model
- Specification of subpopulations for grouping of data for goodness-of-fit tests
- Listing of counts, predicted counts, and residuals by subpopulations
- Correction of variance estimates for over-dispersion
- Covariance matrix of the parameter estimates
- Tests of linear combinations of parameters
- Explicit specification of nested models
- Fit 1-1 matched conditional logistic regression models using differenced variables

Chapter 2

Logistic Regression

Logistic regression is useful for situations in which you want to be able to predict the presence or absence of a characteristic or outcome based on values of a set of predictor variables. It is similar to a linear regression model but is suited to models where the dependent variable is dichotomous. Logistic regression coefficients can be used to estimate odds ratios for each of the independent variables in the model. Logistic regression is applicable to a broader range of research situations than discriminant analysis.

Example. What lifestyle characteristics are risk factors for coronary heart disease (CHD)? Given a sample of patients measured on smoking status, diet, exercise, alcohol use, and CHD status, you could build a model using the four lifestyle variables to predict the presence or absence of CHD in a sample of patients. The model can then be used to derive estimates of the odds ratios for each factor to tell you, for example, how much more likely smokers are to develop CHD than nonsmokers.

Statistics. For each analysis: total cases, selected cases, valid cases. For each categorical variable: parameter coding. For each step: variable(s) entered or removed, iteration history, –2 log-likelihood, goodness of fit, Hosmer-Lemeshow goodness-of-fit statistic, model chi-square, improvement chi-square, classification table, correlations between variables, observed groups and predicted probabilities chart, residual chi-square. For each variable in the equation: coefficient (B), standard error of B, Wald statistic, estimated odds ratio ($\exp(B)$), confidence interval for $\exp(B)$, log-likelihood if term removed from model. For each variable not in the equation: score statistic. For each case: observed group, predicted probability, predicted group, residual, standardized residual.

Methods. You can estimate models using block entry of variables or any of the following stepwise methods: forward conditional, forward LR, forward Wald, backward conditional, backward LR, or backward Wald.

Logistic Regression Data Considerations

Data. The dependent variable should be dichotomous. Independent variables can be interval level or categorical; if categorical, they should be dummy or indicator coded (there is an option in the procedure to recode categorical variables automatically).

Assumptions. Logistic regression does not rely on distributional assumptions in the same sense that discriminant analysis does. However, your solution may be more stable if your predictors have a multivariate normal distribution. Additionally, as with other forms of regression, multicollinearity among the predictors can lead to biased estimates and inflated standard errors. The procedure is most effective when group membership is a truly categorical variable; if group membership is based on values of a continuous variable (for example, "high IQ" versus "low IQ"), you should consider using linear regression to take advantage of the richer information offered by the continuous variable itself.

Related procedures. Use the Scatterplot procedure to screen your data for multicollinearity. If assumptions of multivariate normality and equal variance-covariance matrices are met, you may be able to get a quicker solution using the Discriminant Analysis procedure. If all of your predictor variables are categorical, you can also use the Loglinear procedure. If your dependent variable is continuous, use the Linear Regression procedure. You can use the ROC Curve procedure to plot probabilities saved with the Logistic Regression procedure.

Obtaining a Logistic Regression Analysis

▶ From the menus choose:
Analyze
  Regression
    Binary Logistic...

Logistic Regression

Figure 2-1
Logistic Regression dialog box

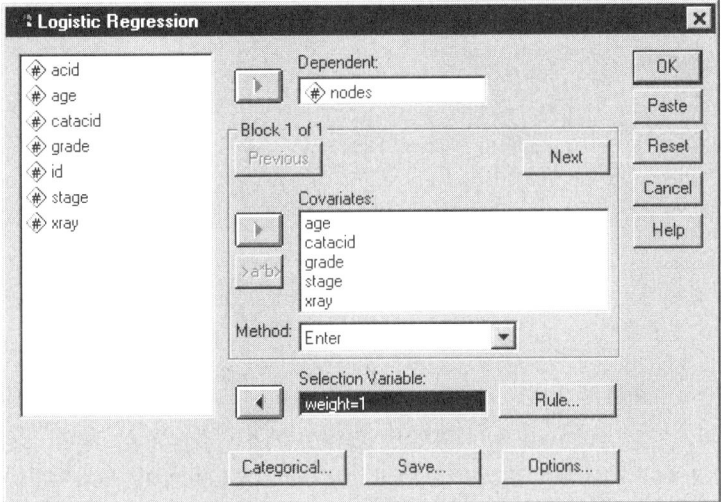

▶ Select one dichotomous dependent variable. This variable may be numeric or short string.

▶ Select one or more covariates. To include interaction terms, select all of the variables involved in the interaction and then select >a*b>.

To enter variables in groups (**blocks**), select the covariates for a block, and click Next to specify a new block. Repeat until all blocks have been specified.

Optionally, you can select cases for analysis. Choose a selection variable, and click Rule.

Chapter 2

Logistic Regression Set Rule

Figure 2-2
Logistic Regression Set Rule dialog box

Cases defined by the selection rule are included in model estimation. For example, if you selected a variable and equals and specified a value of 5, then only the cases for which the selected variable has a value equal to 5 are included in estimating the model.

Statistics and classification results are generated for both selected and unselected cases. This provides a mechanism for classifying new cases based on previously existing data, or for partitioning your data into training and testing subsets, to perform validation on the model generated.

Logistic Regression Variable Selection Methods

Method selection allows you to specify how independent variables are entered into the analysis. Using different methods, you can construct a variety of regression models from the same set of variables.

- **Enter.** A procedure for variable selection in which all variables in a block are entered in a single step.
- **Forward Selection (Conditional).** Stepwise selection method with entry testing based on the significance of the score statistic, and removal testing based on the probability of a likelihood-ratio statistic based on conditional parameter estimates.
- **Forward Selection (Likelihood Ratio).** Stepwise selection method with entry testing based on the significance of the score statistic, and removal testing based on the probability of a likelihood-ratio statistic based on the maximum partial likelihood estimates.
- **Forward Selection (Wald).** Stepwise selection method with entry testing based on the significance of the score statistic, and removal testing based on the probability of the Wald statistic.

- **Backward Elimination (Conditional).** Backward stepwise selection. Removal testing is based on the probability of the likelihood-ratio statistic based on conditional parameter estimates.
- **Backward Elimination (Likelihood Ratio).** Backward stepwise selection. Removal testing is based on the probability of the likelihood-ratio statistic based on the maximum partial likelihood estimates.
- **Backward Elimination (Wald).** Backward stepwise selection. Removal testing is based on the probability of the Wald statistic.

The significance values in your output are based on fitting a single model. Therefore, the significance values are generally invalid when a stepwise method is used.

All independent variables selected are added to a single regression model. However, you can specify different entry methods for different subsets of variables. For example, you can enter one block of variables into the regression model using stepwise selection and a second block using forward selection. To add a second block of variables to the regression model, click Next.

Logistic Regression Define Categorical Variables

Figure 2-3
Logistic Regression Define Categorical Variables dialog box

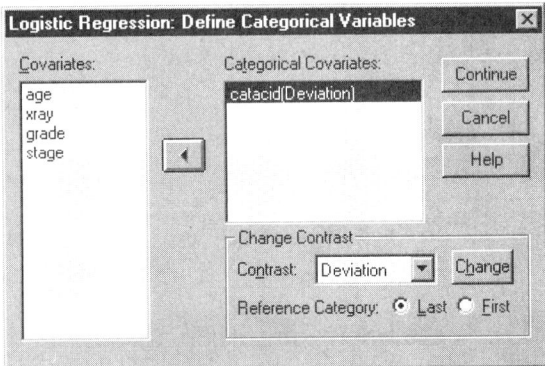

You can specify details of how the Logistic Regression procedure will handle categorical variables:

Covariates. Contains a list of all of the covariates specified in the main dialog box, either by themselves or as part of an interaction, in any layer. If some of these are string variables or are categorical, you can use them only as categorical covariates.

Categorical Covariates. Lists variables identified as categorical. Each variable includes a notation in parentheses indicating the contrast coding to be used. String variables (denoted by the symbol < following their names) are already present in the Categorical Covariates list. Select any other categorical covariates from the Covariates list and move them into the Categorical Covariates list.

Change Contrast. Allows you to change the contrast method. Available contrast methods are:

- **Indicator.** Contrasts indicate the presence or absence of category membership. The reference category is represented in the contrast matrix as a row of zeros.
- **Simple.** Each category of the predictor variable (except the reference category) is compared to the reference category.
- **Difference.** Each category of the predictor variable except the first category is compared to the average effect of previous categories. Also known as reverse Helmert contrasts.
- **Helmert.** Each category of the predictor variable except the last category is compared to the average effect of subsequent categories.
- **Repeated.** Each category of the predictor variable except the first category is compared to the category that precedes it.
- **Polynomial.** Orthogonal polynomial contrasts. Categories are assumed to be equally spaced. Polynomial contrasts are available for numeric variables only.
- **Deviation.** Each category of the predictor variable except the reference category is compared to the overall effect.

If you select Deviation, Simple, or Indicator, select either First or Last as the reference category. Note that the method is not actually changed until you click Change.

String covariates must be categorical covariates. To remove a string variable from the Categorical Covariates list, you must remove all terms containing the variable from the Covariates list in the main dialog box.

Logistic Regression Save New Variables

Figure 2-4
Logistic Regression Save New Variables dialog box

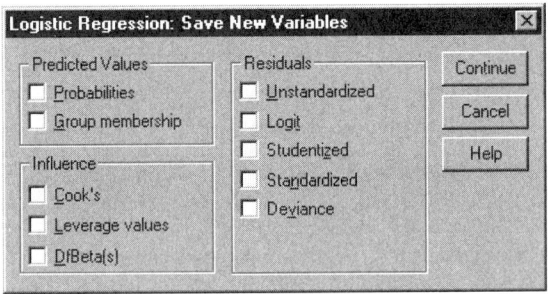

You can save results of the logistic regression as new variables in the working data file:

Predicted Values. Saves values predicted by the model. Available options are Probabilities and Group membership.

- **Probabilities.** For each case, saves the predicted probability of occurrence of the event. A table in the output displays name and contents of any new variables.

- **Predicted Group Membership.** The group with the largest posterior probability, based on discriminant scores. The group the model predicts the case belongs to.

Influence. Saves values from statistics that measure the influence of cases on predicted values. Available options are Cook's, Leverage values, and DfBeta(s).

- **Cook's.** The logistic regression analog of Cook's influence statistic. A measure of how much the residuals of all cases would change if a particular case were excluded from the calculation of the regression coefficients.

- **Leverage Value.** The relative influence of each observation on the model's fit.

- **DfBeta(s).** The difference in beta value is the change in the regression coefficient that results from the exclusion of a particular case. A value is computed for each term in the model, including the constant.

Residuals. Saves residuals. Available options are Unstandardized, Logit, Studentized, Standardized, and Deviance.

- **Unstandardized Residuals.** The difference between an observed value and the value predicted by the model.

- **Logit Residual.** The residual for the case if it is predicted in the logit scale. The logit residual is the residual divided by the predicted probability times 1 minus the predicted probability.
- **Studentized Residual.** The change in the model deviance if a case is excluded.
- **Standardized Residuals.** The residual divided by an estimate of its standard deviation. Standardized residuals which are also known as Pearson residuals, have a mean of 0 and a standard deviation of 1.
- **Deviance.** Residuals based on the model deviance.

Logistic Regression Options

Figure 2-5
Logistic Regression Options dialog box

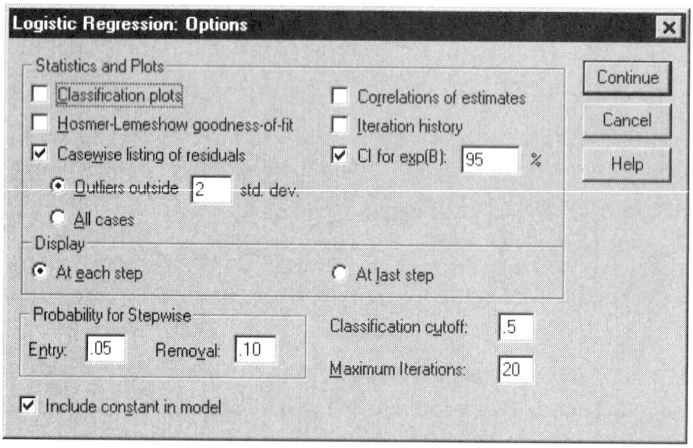

You can specify options for your logistic regression analysis:

Statistics and Plots. Allows you to request statistics and plots. Available options are Classification plots, Hosmer-Lemeshow goodness-of-fit, Casewise listing of residuals, Correlations of estimates, Iteration history, and CI for exp(*B*). Select one of the alternatives in the Display group to display statistics and plots either At each step or, only for the final model, At last step.

- **Hosmer-Lemeshow goodness-of-fit statistic.** This goodness-of-fit statistic is more robust than the traditional goodness-of-fit statistic used in logistic regression, particularly for models with continuous covariates and studies with small sample

Logistic Regression

sizes. It is based on grouping cases into deciles of risk and comparing the observed probability with the expected probability within each decile.

Probability for Stepwise. Allows you to control the criteria by which variables are entered into and removed from the equation. You can specify criteria for Entry or Removal of variables.

- **Probability for Stepwise.** A variable is entered into the model if the probability of its score statistic is less than the Entry value, and is removed if the probability is greater than the Removal value. To override the default settings, enter positive values for Entry and Removal. Entry must be less than Removal.

Classification cutoff. Allows you to determine the cut point for classifying cases. Cases with predicted values that exceed the classification cutoff are classified as positive, while those with predicted values smaller than the cutoff are classified as negative. To change the default, enter a value between 0.01 and 0.99.

Maximum Iterations. Allows you to change the maximum number of times that the model iterates before terminating.

Include constant in model. Allows you to indicate whether the model should include a constant term. If disabled, the constant term will equal 0.

LOGISTIC REGRESSION Command Additional Features

The SPSS command language also allows you to:

- Identify casewise output by the values or variable labels of a variable.
- Control the spacing of iteration reports. Rather than printing parameter estimates after every iteration, you can request parameter estimates after every *n*th iteration.
- Change the criteria for terminating iteration and checking for redundancy.
- Specify a variable list for casewise listings.
- Conserve memory by holding the data for each split file group in an external scratch file during processing.

Chapter 3

Multinomial Logistic Regression

Multinomial Logistic Regression is useful for situations in which you want to be able to classify subjects based on values of a set of predictor variables. This type of regression is similar to logistic regression, but it is more general because the dependent variable is not restricted to two categories.

Example. In order to market films more effectively, movie studios want to predict what type of film a moviegoer is likely to see. By performing a Multinomial Logistic Regression, the studio can determine the strength of influence a person's age, gender, and dating status has upon the type of film they prefer. The studio can then slant the advertising campaign of a particular movie toward a group of people likely to go see it.

Statistics. Iteration history, parameter coefficients, asymptotic covariance and correlation matrices, likelihood-ratio tests for model and partial effects, –2 log-likelihood. Pearson and deviance chi-square goodness of fit. Cox and Snell, Nagelkerke, and McFadden R^2. Classification: observed versus predicted frequencies by response category. Crosstabulation: observed and predicted frequencies (with residuals) and proportions by covariate pattern and response category.

Methods. A multinomial logit model is fit for the full factorial model or a user-specified model. Parameter estimation is performed through an iterative maximum-likelihood algorithm.

Multinomial Logistic Regression Data Considerations

Data. The dependent variable should be categorical. Independent variables can be factors or covariates. In general, factors should be categorical variables and covariates should be continuous variables.

Assumptions. It is assumed that the odds ratio of any two categories are independent of all other response categories. For example, if a new product is introduced to a market, this assumption states that the market shares of all other products are affected proportionally equally. Also, given a covariate pattern, the responses are assumed to be independent multinomial variables.

Obtaining a Multinomial Logistic Regression

▶ From the menus choose:

Analyze
  Regression
    Multinomial Logistic...

Figure 3-1
Multinomial Logistic Regression dialog box

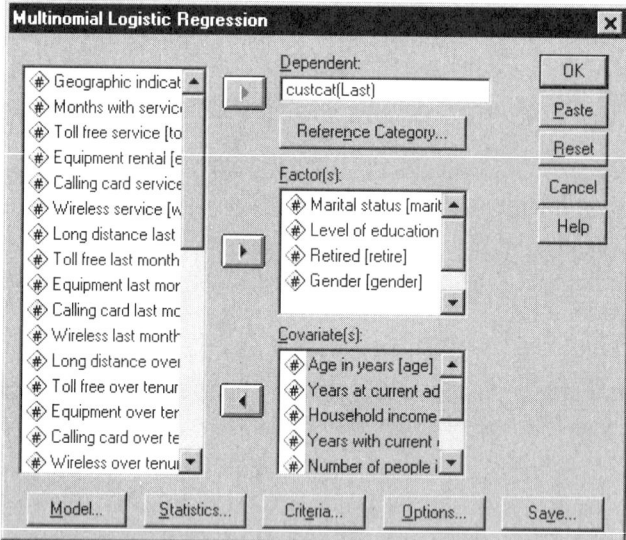

▶ Select one dependent variable.

▶ Factors are optional and can be either numeric or categorical.

▶ Covariates are optional but must be numeric if specified.

Multinomial Logistic Regression Models

Figure 3-2
Multinomial Logistic Regression Model dialog box

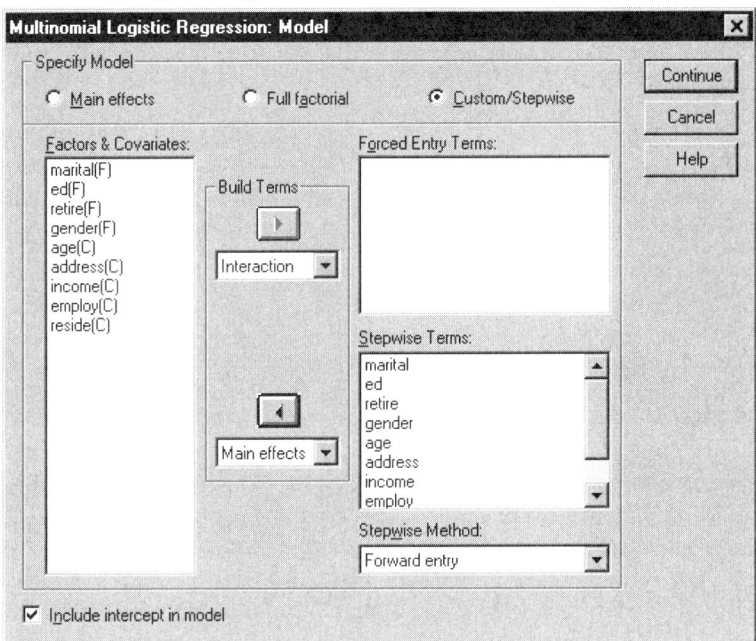

By default, the Multinomial Logistic Regression procedure produces a model with the factor and covariate main effects, but you can specify a custom model or request stepwise model selection with this dialog box.

Specify Model. A main-effects model contains the covariate and factor main effects but no interaction effects. A full factorial model contains all main effects and all factor-by-factor interactions. It does not contain covariate interactions. You can create a custom model to specify subsets of factor interactions or covariate interactions, or request stepwise selection of model terms.

Factors and Covariates. The factors and covariates are listed with (F) for factor and (C) for covariate.

Forced Entry Terms. Terms added to the forced entry list are always included in the model.

Stepwise Terms. Terms added to the stepwise list are included in the model according to one of the following user-selected methods:

- **Forward entry.** This method begins with no stepwise terms in the model. At each step, the most significant term is added to the model until none of the stepwise terms left out of the model would have a statistically significant contribution if added to the model.

- **Backward elimination.** This method begins by entering all terms specified on the stepwise list into the model. At each step, the least significant stepwise term is removed from the model until all of the remaining stepwise terms have a statistically significant contribution to the model.

- **Forward stepwise.** This method begins with the model that would be selected by the forward entry method. From there, the algorithm alternates between backward elimination on the stepwise terms in the model and forward entry on the terms left out of the model. This continues until no terms meet the entry or removal criteria.

- **Backward stepwise.** This method begins with the model that would be selected by the backward elimination method. From there, the algorithm alternates between forward entry on the terms left out of the model and backward elimination on the stepwise terms in the model. This continues until no terms meet the entry or removal criteria.

Include intercept in model. Allows you to include or exclude an intercept term for the model.

Build Terms

For the selected factors and covariates:

Interaction. Creates the highest-level interaction term of all selected variables.

Main effects. Creates a main-effects term for each variable selected.

All 2-way. Creates all possible two-way interactions of the selected variables.

All 3-way. Creates all possible three-way interactions of the selected variables.

All 4-way. Creates all possible four-way interactions of the selected variables.

All 5-way. Creates all possible five-way interactions of the selected variables.

Multinomial Logistic Regression Reference Category

Figure 3-3
Multinomial Logistic Regression Reference Category dialog box

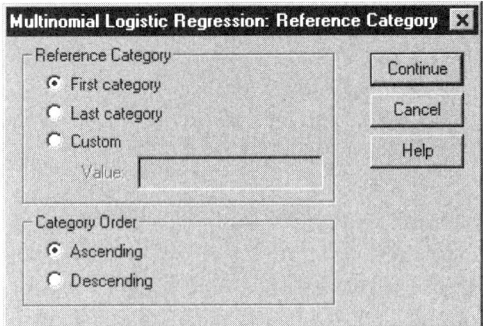

By default, the Multinomial Logistic Regression procedure makes the last category the reference category. This dialog box gives you control of the reference category and the way in which categories are ordered.

Reference Category. Specify the first, last, or a custom category.

Category Order. In ascending order, the lowest value defines the first category and the highest value defines the last. In descending order, the highest value defines the first category and the lowest value defines the last.

Chapter 3

Multinomial Logistic Regression Statistics

Figure 3-4
Multinomial Logistic Regression Statistics dialog box

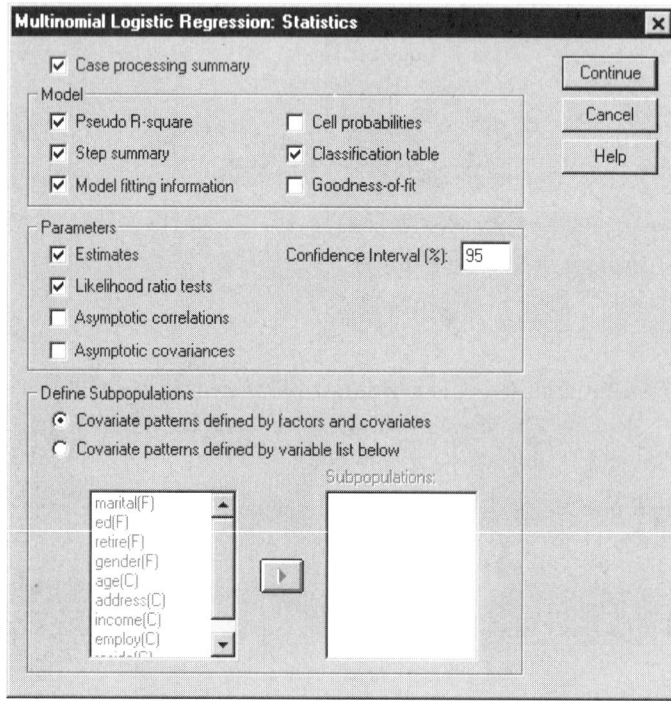

You can specify the following statistics for your Multinomial Logistic Regression:

Case processing summary. This table contains information about the specified categorical variables.

Model. Statistics for the overall model.

- **Summary statistics.** Prints the Cox and Snell, Nagelkerke, and McFadden R^2 statistics.
- **Step summary.** This table summarizes the effects entered or removed at each step in a stepwise method. It is not produced unless a stepwise model is specified in the Model dialog box.
- **Model fitting information.** This table compares the fitted and intercept-only or null models.

Multinomial Logistic Regression

- **Cell probabilities.** Prints a table of the observed and expected frequencies (with residual) and proportions by covariate pattern and response category.
- **Classification table.** Prints a table of the observed versus predicted responses.
- **Goodness of fit chi-square statistics.** Prints Pearson and likelihood-ratio chi-square statistics. Statistics are computed for the covariate patterns determined by all factors and covariates or by a user-defined subset of the factors and covariates.

Parameters. Statistics related to the model parameters.

- **Estimates.** Prints estimates of the model parameters, with a user-specified level of confidence.
- **Likelihood ratio test.** Prints likelihood-ratio tests for the model partial effects. The test for the overall model is printed automatically.
- **Asymptotic correlations.** Prints matrix of parameter estimate correlations.
- **Asymptotic covariances.** Prints matrix of parameter estimate covariances.

Define Subpopulations. Allows you to select a subset of the factors and covariates in order to define the covariate patterns used by cell probabilities and the goodness-of-fit tests.

Multinomial Logistic Regression Criteria

Figure 3-5
Multinomial Logistic Regression Convergence Criteria dialog box

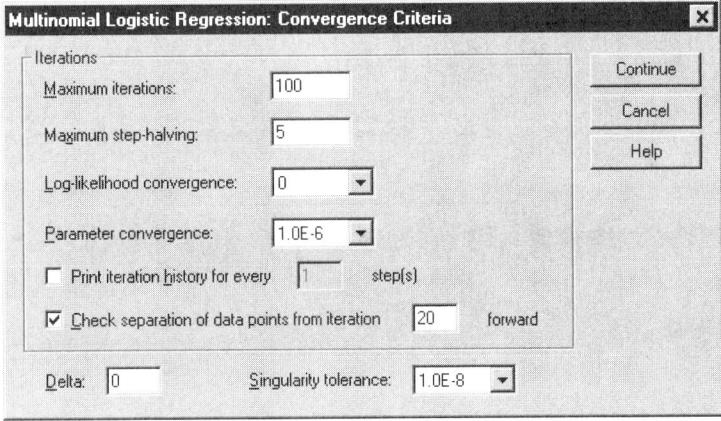

You can specify the following criteria for your Multinomial Logistic Regression:

Iterations. Allows you to specify the maximum number of times you want to cycle through the algorithm, the maximum number of steps in the step-halving, the convergence tolerances for changes in the log-likelihood and parameters, how often the progress of the iterative algorithm is printed, and at what iteration the procedure should begin checking for complete or quasi-complete separation of the data.

- **Log-likelihood Convergence.** Convergence is assumed if the relative change in the log-likelihood function is less than the specified value. The value must be positive.
- **Parameter Convergence.** The algorithm is assumed to have reach the correct estimates if the absolute change or relative change in the parameter estimates is less than this value. The criterion is not used if the value is 0.

Delta. Allows you to specify a non-negative value less than 1. This value is added to each empty cell of the crosstabulation of response category by covariate pattern. This helps to stabilize the algorithm and prevent bias in the estimates.

Singularity tolerance. Allows you to specify the tolerance used in checking for singularities.

Multinomial Logistic Regression Options

Figure 3-6
Multinomial Logistic Regression Options dialog box

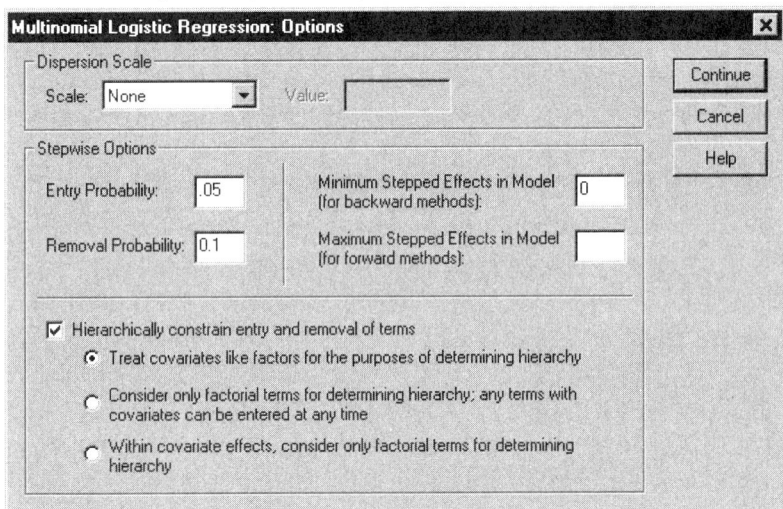

You can specify the following options for your Multinomial Logistic Regression:

Dispersion Scale. Allows you to specify the dispersion scaling value that will be used to correct the estimate of the parameter covariance matrix. Deviance estimates the scaling value using the deviance function (likelihood-ratio chi-square) statistic. Pearson estimates the scaling value using the Pearson chi-square statistic. You can also specify your own scaling value. It must be a positive numeric value.

Stepwise Options. These options give you control of the statistical criteria when stepwise methods are used to build a model. They are ignored unless a stepwise model is specified in the Model dialog box.

- **Entry Probability.** This is the probability of the likelihood-ratio statistic for variable entry. The larger the specified probability, the easier it is for a variable to enter the model. This criterion is ignored unless the forward entry, forward stepwise, or backward stepwise method is selected.

- **Removal Probability.** This is the probability of the likelihood-ratio statistic for variable removal. The larger the specified probability, the easier it is for a variable to remain in the model. This criterion is ignored unless the backward elimination, forward stepwise, or backward stepwise method is selected.

- **Minimum Stepped Effect in Model.** When using the backward elimination or backward stepwise methods, this specifies the minimum number of terms to include in the model. The intercept is not counted as a model term.

- **Maximum Stepped Effect in Model.** When using the forward entry or forward stepwise methods, this specifies the maximum number of terms to include in the model. The intercept is not counted as a model term.

- **Hierarchically constrain entry and removal of terms.** This option allows you to choose whether to place restrictions on the inclusion of model terms. Hierarchy requires that for any term to be included, all lower order terms that are a part of the term to be included must be in the model first. For example, if the hierarchy requirement is in effect, the factors *Marital status* and *Gender* must both be in the model before the *Marital Status*Gender* interaction can be added. The three radio button options determine the role of covariates in determining hierarchy.

Multinomial Logistic Regression Save

Figure 3-7
Multinomial Logistic Regression Save dialog box

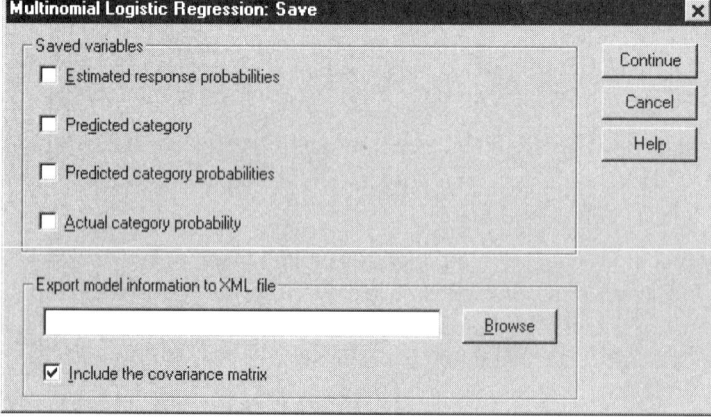

Multinomial Logistic Regression

The Save dialog box allows you to save variables to the working file and export model information to an external file.

Saved variables:

- **Estimated response probabilities.** These are the estimated probabilities of classifying a factor/covariate pattern into the response categories. There are as many estimated probabilities as there are categories of the response variable; up to 25 will be saved.
- **Predicted category.** This is the response category with the largest expected probability for a factor/covariate pattern.
- **Predicted category probabilities.** This is the maximum of the estimated response probabilities.
- **Actual category probability.** This is the estimated probability of classifying a factor/covariate pattern into the observed category.

Export model information to XML file. Parameter estimates and (optionally) their covariances are exported to the specified file. *SmartScore* and future releases of *WhatIf?* will be able to use this file.

NOMREG Command Additional Features

The SPSS command language also allows you to:

- Specify the reference category of the dependent variable.
- Include cases with user-missing values.
- Customize hypothesis tests by specifying null hypotheses as linear combinations of parameters.

See the *SPSS Command Syntax Reference* for complete syntax information.

Chapter 4

Probit Analysis

This procedure measures the relationship between the strength of a stimulus and the proportion of cases exhibiting a certain response to the stimulus. It is useful for situations where you have a dichotomous output that is thought to be influenced or caused by levels of some independent variable(s) and is particularly well suited to experimental data. This procedure will allow you to estimate the strength of a stimulus required to induce a certain proportion of responses, such as the median effective dose.

Example. How effective is a new pesticide at killing ants, and what is an appropriate concentration to use? You might perform an experiment in which you expose samples of ants to different concentrations of the pesticide and then record the number of ants killed and the number of ants exposed. Applying probit analysis to these data, you can determine the strength of the relationship between concentration and killing, and you can determine what the appropriate concentration of pesticide would be if you wanted to be sure to kill, say, 95% of exposed ants.

Statistics. Regression coefficients and standard errors, intercept and standard error, Pearson goodness-of-fit chi-square, observed and expected frequencies, and confidence intervals for effective levels of independent variable(s). Plots: transformed response plots.

This procedure uses the algorithms proposed and implemented in NPSOL® by Gill, Murray, Saunders & Wright to estimate the model parameters.

Probit Analysis Data Considerations

Data. For each value of the independent variable (or each combination of values for multiple independent variables), your response variable should be a count of the number of cases with those values that show the response of interest, and the total observed variable should be a count of the total number of cases with those values for the independent variable. The factor variable should be categorical, coded as integers.

Assumptions. Observations should be independent. If you have a large number of values for the independent variables relative to the number of observations, as you might in an observational study, the chi-square and goodness-of-fit statistics may not be valid.

Related procedures. Probit analysis is closely related to logistic regression; in fact, if you choose the logit transformation, this procedure will essentially compute a logistic regression. In general, probit analysis is appropriate for designed experiments, whereas logistic regression is more appropriate for observational studies. The differences in output reflect these different emphases. The probit analysis procedure reports estimates of effective values for various rates of response (including median effective dose), while the logistic regression procedure reports estimates of odds ratios for independent variables.

Obtaining a Probit Analysis

▶ From the menus choose:
Analyze
  Regression
    Probit...

Figure 4-1
Probit Analysis dialog box

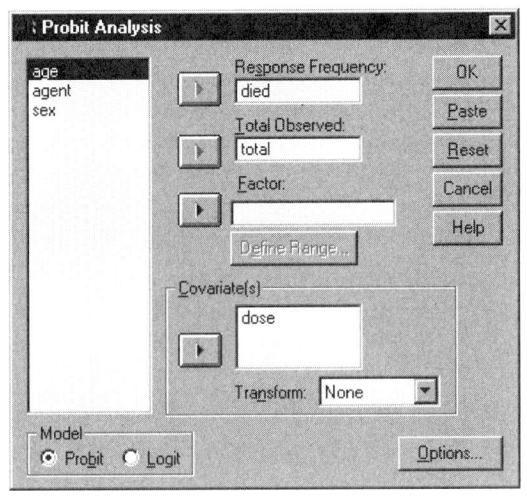

▶ Select a response frequency variable. This variable indicates the number of cases exhibiting a response to the test stimulus. The values of this variable cannot be negative.

▶ Select a total observed variable. This variable indicates the number of cases to which the stimulus was applied. The values of this variable cannot be negative and cannot be less than the values of the response frequency variable for each case.

Optionally, you can select a Factor variable. If you do, click Define Range to define the groups.

▶ Select one or more covariate(s). This variable contains the level of the stimulus applied to each observation. If you want to transform the covariate, select a transformation from the Transform drop-down list. If no transformation is applied and there is a control group, then the control group is included in the analysis.

▶ Select either the Probit or Logit model.
  - **Probit Model.** Applies the probit transformation (the inverse of the cumulative standard normal distribution function) to the response proportions.
  - **Logit Model.** Applies the logit (log odds) transformation to the response proportions.

Chapter 4

Probit Analysis Define Range

Figure 4-2
Probit Analysis Define Range dialog box

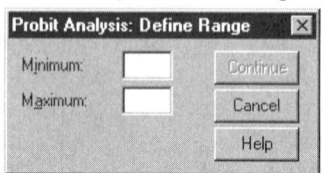

This allows you to specify the levels of the factor variable that will be analyzed. The factor levels must be coded as consecutive integers, and all levels in the range that you specify will be analyzed.

Probit Analysis Options

Figure 4-3
Probit Analysis Options dialog box

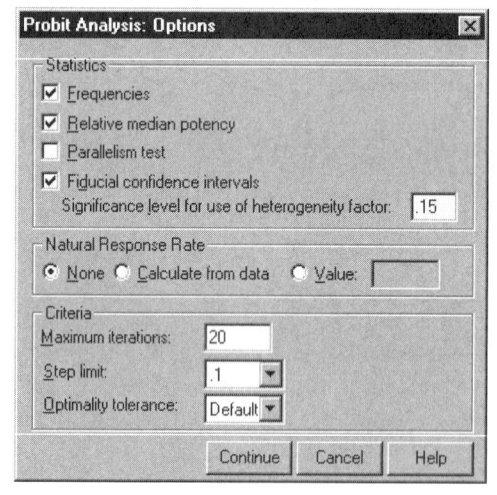

You can specify options for your probit analysis:

Statistics. Allows you to request the following optional statistics: Frequencies, Relative median potency, Parallelism test, and Fiducial confidence intervals.

- **Relative Median Potency.** Displays the ratio of median potencies for each pair of factor levels. Also shows 95% confidence limits for each relative median potency. Relative median potencies are not available if you do not have a factor variable or if you have more than one covariate.
- **Parallelism Test.** A test of the hypothesis that all factor levels have a common slope.
- **Fiducial Confidence Intervals.** Confidence intervals for the dosage of agent required to produce a certain probability of response.

Fiducial confidence intervals and Relative median potency are unavailable if you have selected more than one covariate. Relative median potency and Parallelism test are available only if you have selected a factor variable.

Natural Response Rate. Allows you to indicate a natural response rate even in the absence of the stimulus. Available alternatives are None, Calculate from data, or Value.

- **Calculate from Data.** Estimate the natural response rate from the sample data. Your data should contain a case representing the control level, for which the value of the covariate(s) is 0. Probit estimates the natural response rate using the proportion of responses for the control level as an initial value.
- **Value.** Sets the natural response rate in the model (select this item when you know the natural response rate in advance). Enter the natural response proportion (the proportion must be less than 1). For example, if the response occurs 10% of the time when the stimulus is 0, enter 0.10.

Criteria. Allows you to control parameters of the iterative parameter-estimation algorithm. You can override the defaults for Maximum iterations, Step limit, and Optimality tolerance.

PROBIT Command Additional Features

The SPSS command language also allows you to:
- Request an analysis on both the probit and logit models.
- Control the treatment of missing values.
- Transform the covariates by bases other than base 10 or natural log.

Chapter 5

Nonlinear Regression

Nonlinear regression is a method of finding a nonlinear model of the relationship between the dependent variable and a set of independent variables. Unlike traditional linear regression, which is restricted to estimating linear models, nonlinear regression can estimate models with arbitrary relationships between independent and dependent variables. This is accomplished using iterative estimation algorithms. Note that this procedure is not necessary for simple polynomial models of the form $Y = A + BX^{**}2$. By defining $W = X^{**}2$, we get a simple linear model, $Y = A + BW$, which can be estimated using traditional methods such as the Linear Regression procedure.

Example. Can population be predicted based on time? A scatterplot shows that there seems to be a strong relationship between population and time, but the relationship is nonlinear, so it requires the special estimation methods of the Nonlinear Regression procedure. By setting up an appropriate equation, such as a logistic population growth model, we can get a good estimate of the model, allowing us to make predictions about population for times that were not actually measured.

Statistics. For each iteration: parameter estimates and residual sum of squares. For each model: sum of squares for regression, residual, uncorrected total and corrected total, parameter estimates, asymptotic standard errors, and asymptotic correlation matrix of parameter estimates.

Note: Constrained nonlinear regression uses the algorithms proposed and implemented in NPSOL® by Gill, Murray, Saunders, and Wright to estimate the model parameters.

Chapter 5

Nonlinear Regression Data Considerations

Data. The dependent and independent variables should be quantitative. Categorical variables, such as religion, major, or region of residence, need to be recoded to binary (dummy) variables or other types of contrast variables.

Assumptions. Results are valid only if you have specified a function that accurately describes the relationship between dependent and independent variables. Additionally, the choice of good starting values is very important. Even if you've specified the correct functional form of the model, if you use poor starting values, your model may fail to converge or you may get a locally optimal solution rather than one that is globally optimal.

Related procedures. Many models that appear nonlinear at first can be transformed to a linear model, which can be analyzed using the Linear Regression procedure. If you are uncertain what the proper model should be, the Curve Estimation procedure can help to identify useful functional relations in your data.

Obtaining a Nonlinear Regression Analysis

▶ From the menus choose:
Analyze
  Regression
    Nonlinear...

Nonlinear Regression

Figure 5-1
Nonlinear Regression dialog box

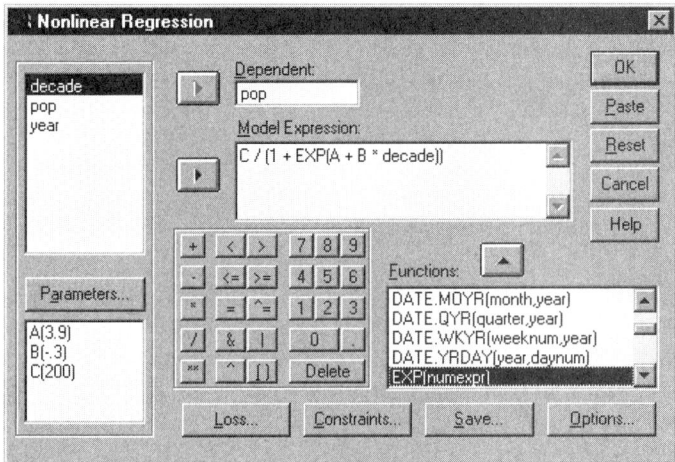

▶ Select one numeric dependent variable from the list of variables in your working data file.

▶ To build a model expression, enter the expression in the Model field or paste components (variables, parameters, functions) into the field.

▶ Identify parameters in your model by clicking Parameters.

A segmented model (one that takes different forms in different parts of its domain) must be specified by using conditional logic within the single model statement.

Conditional Logic (Nonlinear Regression)

You can specify a segmented model using conditional logic. To use conditional logic within a model expression or a loss function, you form the sum of a series of terms, one for each condition. Each term consists of a logical expression (in parentheses) multiplied by the expression that should result when that logical expression is true.

For example, consider a segmented model that equals 0 for X<=0, X for 0<X<1, and 1 for X>=1. The expression for this is:

(X<=0)*0 + (X>0 & X < 1)*X + (X>=1)*1.

The logical expressions in parentheses all evaluate to 1 (true) or 0 (false). Therefore:

If X<=0, the above reduces to 1*0 + 0*X + 0*1=0.

If 0<X<1, it reduces to 0*0 + 1*X +0*1 = X.

If X>=1, it reduces to 0*0 + 0*X + 1*1 = 1.

More complicated examples can be easily built by substituting different logical expressions and outcome expressions. Remember that double inequalities, such as 0<X<1, must be written as compound expressions, such as (X>0 & X < 1).

String variables can be used within logical expressions:

(city='New York')*costliv + (city='Des Moines')*0.59*costliv

This yields one expression (the value of the variable *costliv*) for New Yorkers and another (59% of that value) for Des Moines residents. String constants must be enclosed in quotation marks or apostrophes, as shown here.

Nonlinear Regression Parameters

Figure 5-2
Nonlinear Regression Parameters dialog box

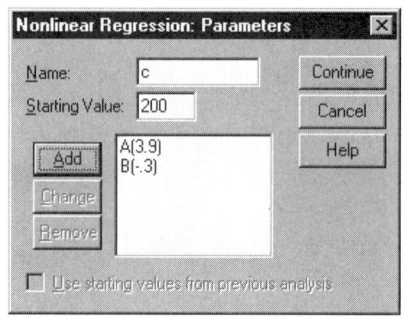

Parameters are the parts of your model that the Nonlinear Regression procedure estimates. Parameters can be additive constants, multiplicative coefficients, exponents, or values used in evaluating functions. All parameters that you have defined will appear (with their initial values) on the Parameters list in the main dialog box.

Nonlinear Regression

Name. You must specify a name for each parameter. This name must be a valid SPSS variable name and must be the name used in the model expression in the main dialog box.

Starting Value. Allows you to specify a starting value for the parameter, preferably as close as possible to the expected final solution. Poor starting values can result in failure to converge or in convergence on a solution that is local (rather than global) or is physically impossible.

Use starting values from previous analysis. If you have already run a nonlinear regression from this dialog box, you can select this option to obtain the initial values of parameters from their values in the previous run. This permits you to continue searching when the algorithm is converging slowly. (The initial starting values will still appear on the Parameters list in the main dialog box.)

Note: This selection persists in this dialog box for the rest of your session. If you change the model, be sure to deselect it.

Nonlinear Regression Common Models

The table below provides example model syntax for many published nonlinear regression models. A model selected at random is not likely to fit your data well. Appropriate starting values for the parameters are necessary, and some models require constraints in order to converge.

Table 5-1
Example model syntax

Name	Model expression
Asymptotic Regression	b1 + b2 *exp(b3 * x)
Asymptotic Regression	b1 −(b2 *(b3 ** x))
Density	(b1 + b2 * x)**(−1/ b3)
Gauss	b1 *(1− b3 *exp(−b2 * x **2))
Gompertz	b1 *exp(−b2 * exp(−b3 * x))
Johnson-Schumacher	b1 *exp(−b2 / (x + b3))
Log-Modified	(b1 + b3 * x) ** b2
Log-Logistic	b1 −ln(1+ b2 *exp(−b3 * x))

Name	Model expression
Metcherlich Law of Diminishing Returns	b1 + b2 *exp(–b3 * x)
Michaelis Menten	b1* x /(x + b2)
Morgan-Mercer-Florin	(b1 * b2 + b3 * x ** b4)/(b2 + x ** b4)
Peal-Reed	b1 /(1+ b2 *exp(–(b3 * x + b4 * x **2+ b5 * x **3)))
Ratio of Cubics	(b1 + b2 * x + b3 * x **2+ b4 * x **3)/(b5 * x **3)
Ratio of Quadratics	(b1 + b2 * x + b3 * x **2)/(b4 * x **2)
Richards	b1 /((1+ b3 *exp(– b2 * x))**(1/ b4))
Verhulst	b1 /(1 + b3 * exp(– b2 * x))
Von Bertalanffy	(b1 ** (1 – b4) – b2 * exp(–b3 * x)) ** (1/(1 –b4))
Weibull	b1 – b2 *exp(– b3 * x ** b4)
Yield Density	(b1 + b2 * x + b3 * x **2)**(–1)

Nonlinear Regression Loss Function

Figure 5-3
Nonlinear Regression Loss Function dialog box

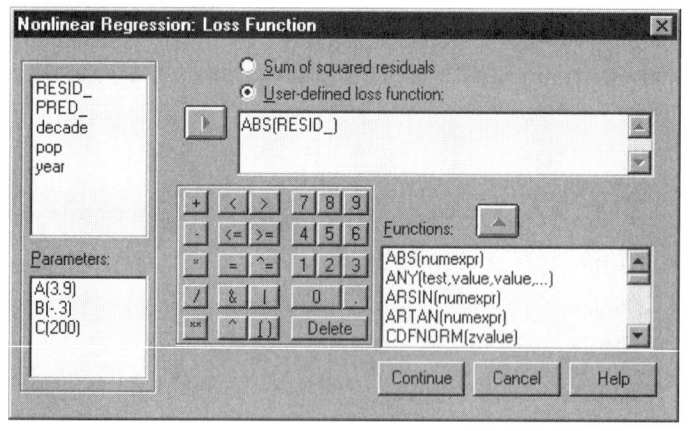

The **loss function** in nonlinear regression is the function that is minimized by the algorithm. Select either Sum of squared residuals to minimize the sum of the squared residuals or User-defined loss function to minimize a different function.

Nonlinear Regression

If you select User-defined loss function, you must define the loss function whose sum (across all cases) should be minimized by the choice of parameter values.

- Most loss functions involve the special variable *RESID_*, which represents the residual. (The default Sum of squared residuals loss function could be entered explicitly as RESID_**2.) If you need to use the predicted value in your loss function, it is equal to the dependent variable minus the residual.
- It is possible to specify a conditional loss function using conditional logic.

You can either type an expression in the User-defined loss function field or paste components of the expression into the field. String constants must be enclosed in quotation marks or apostrophes, and numeric constants must be typed in American format, with the dot as a decimal delimiter.

Nonlinear Regression Parameter Constraints

Figure 5-4
Nonlinear Regression Parameter Constraints dialog box

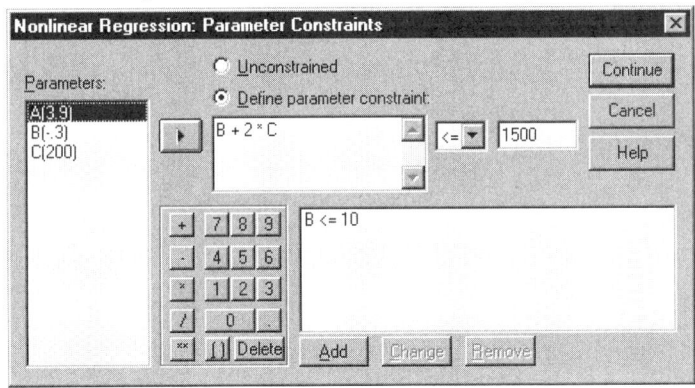

A **constraint** is a restriction on the allowable values for a parameter during the iterative search for a solution. Linear expressions are evaluated before a step is taken, so you can use linear constraints to prevent steps that might result in overflows. Nonlinear expressions are evaluated after a step is taken.

Each equation or inequality requires the following elements:

- An expression involving at least one parameter in the model. Type the expression or use the keypad, which allows you to paste numbers, operators, or parentheses into the expression. You can either type in the required parameter(s) along with the rest of the expression or paste from the Parameters list at the left. You cannot use ordinary variables in a constraint.
- One of the three logical operators <=, =, or >=.
- A numeric constant, to which the expression is compared using the logical operator. Type the constant. Numeric constants must be typed in American format, with the dot as a decimal delimiter.

Nonlinear Regression Save New Variables

Figure 5-5
Nonlinear Regression Save New Variables dialog box

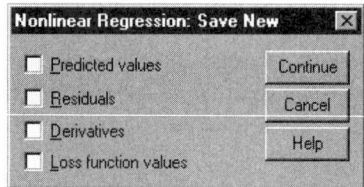

You can save a number of new variables to your active data file. Available options are Predicted values, Residuals, Derivatives, and Loss function values. These variables can be used in subsequent analyses to test the fit of the model or to identify problem cases.

- **Predicted Values.** Saves predicted values with the variable name pred_.
- **Residuals.** Saves residuals with the variable name resid.
- **Derivatives.** One derivative is saved for each model parameter. Derivative names are created by prefixing 'd.' to the first six characters of parameter names.
- **Loss Function Values.** This option is available if you specify your own loss function. The variable name loss_ is assigned to the values of the loss function.

Nonlinear Regression Options

Figure 5-6
Nonlinear Regression Options dialog box

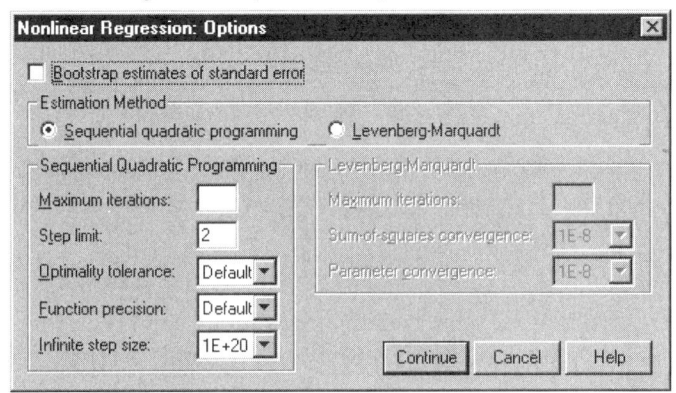

Options allow you to control various aspects of your nonlinear regression analysis:

Bootstrap Estimates. A method of estimating the standard error of a statistic using repeated samples from the original data set. This is done by sampling (with replacement) to get many samples of the same size as the original data set. The nonlinear equation is estimated for each of these samples. The standard error of each parameter estimate is then calculated as the standard deviation of the bootstrapped estimates. Parameter values from the original data are used as starting values for each bootstrap sample. This requires the sequential quadratic programming algorithm.

Estimation Method. Allows you to select an estimation method, if possible. (Certain choices in this or other dialog boxes require the sequential quadratic programming algorithm.) Available alternatives include Sequential quadratic programming and Levenberg-Marquardt.

- **Sequential Quadratic Programming.** This method is available for constrained and unconstrained models. Sequential quadratic programming is used automatically if you specify a constrained model, a user-defined loss function, or bootstrapping. You can enter new values for Maximum iterations and Step limit, and you can change the selection in the drop-down lists for Optimality tolerance, Function precision, and Infinite step size.

- **Levenberg-Marquardt.** This is the default algorithm for unconstrained models. The Levenberg-Marquardt method is not available if you specify a constrained model, a user-defined loss function, or bootstrapping. You can enter new values for Maximum iterations, and you can change the selection in the drop-down lists for Sum-of-squares convergence and Parameter convergence.

Interpreting Nonlinear Regression Results

Nonlinear regression problems often present computational difficulties:

- The choice of initial values for the parameters influences convergence. Try to choose initial values that are reasonable and, if possible, close to the expected final solution.

- Sometimes one algorithm performs better than the other on a particular problem. In the Options dialog box, select the other algorithm if it is available. (If you specify a loss function or certain types of constraints, you cannot use the Levenberg-Marquardt algorithm.)

- When iteration stops only because the maximum number of iterations has occurred, the "final" model is probably not a good solution. Select Use starting values from previous analysis in the Parameters dialog box to continue the iteration or, better yet, choose different initial values.

- Models that require exponentiation of or by large data values can cause overflows or underflows (numbers too large or too small for the computer to represent). Sometimes you can avoid these by suitable choice of initial values or by imposing constraints on the parameters.

NLR Command Additional Features

The SPSS command language also allows you to:

- Name a file from which to read initial values for parameter estimates.
- Specify more than one model statement and loss function. This makes it easier to specify a segmented model.
- Supply your own derivatives rather than use those calculated by the program.

- Specify the number of bootstrap samples to generate.
- Specify additional iteration criteria, including setting a critical value for derivative checking and defining a convergence criterion for the correlation between the residuals and the derivatives.

Additional criteria for the CNLR (constrained nonlinear regression) command allow you to:

- Specify the maximum number of minor iterations allowed within each major iteration.
- Set a critical value for derivative checking.
- Set a step limit.
- Specify a crash tolerance to determine if initial values are within their specified bounds.

Chapter 6

Weight Estimation

Standard linear regression models assume that variance is constant within the population under study. When this is not the case—for example, when cases that are high on some attribute show more variability than cases that are low on that attribute—linear regression using ordinary least squares (OLS) no longer provides optimal model estimates. If the differences in variability can be predicted from another variable, the Weight Estimation procedure can compute the coefficients of a linear regression model using weighted least squares (WLS), such that the more precise observations (that is, those with less variability) are given greater weight in determining the regression coefficients. The Weight Estimation procedure tests a range of weight transformations and indicates which will give the best fit to the data.

Example. What are the effects of inflation and unemployment on changes in stock prices? Because stocks with higher share values often show more variability than those with low share values, ordinary least squares will not produce optimal estimates. Weight estimation allows you to account for the effect of share price on the variability of price changes in calculating the linear model.

Statistics. Log-likelihood values for each power of the weight source variable tested, multiple R, R-squared, adjusted R-squared, ANOVA table for WLS model, unstandardized and standardized parameter estimates, and log-likelihood for the WLS model.

Weight Estimation Data Considerations

Data. The dependent and independent variables should be quantitative. Categorical variables, such as religion, major, or region of residence, need to be recoded to binary (dummy) variables or other types of contrast variables. The weight variable should be quantitative and should be related to the variability in the dependent variable.

Assumptions. For each value of the independent variable, the distribution of the dependent variable must be normal. The relationship between the dependent variable and each independent variable should be linear, and all observations should be independent. The variance of the dependent variable can vary across levels of the independent variable(s), but the differences must be predictable based on the weight variable.

Related procedures. The Explore procedure can be used to screen your data. Explore provides tests for normality and homogeneity of variance, as well as graphical displays. If your dependent variable seems to have equal variance across levels of independent variables, you can use the Linear Regression procedure. If your data appear to violate an assumption (such as normality), try transforming them. If your data are not related linearly and a transformation does not help, use an alternate model in the Curve Estimation procedure. If your dependent variable is dichotomous—for example, whether a particular sale is completed or whether an item is defective—use the Logistic Regression procedure. If your dependent variable is censored—for example, survival time after surgery—use Life Tables, Kaplan-Meier, or Cox Regression, available in the SPSS Advanced Models option. If your data are not independent—for example, if you observe the same person under several conditions—use the Repeated Measures procedure, available in the SPSS Advanced Models option.

Obtaining a Weight Estimation Analysis

▶ From the menus choose:

Analyze
　Regression
　　Weight Estimation...

Weight Estimation

Figure 6-1
Weight Estimation dialog box

[Weight Estimation dialog box screenshot]

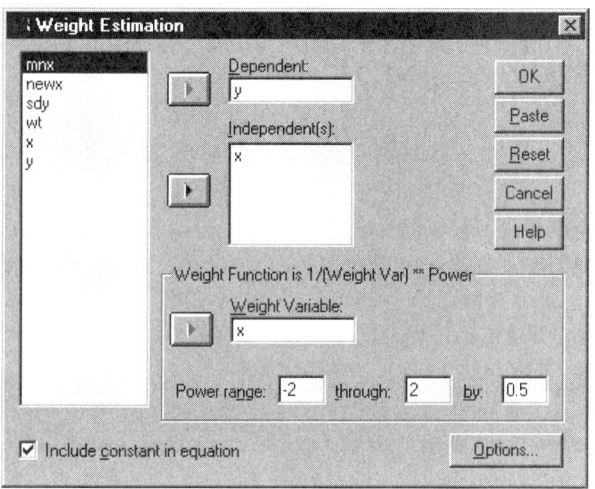

▶ Select one dependent variable.

▶ Select one or more independent variables.

▶ Select the variable that is the source of heteroscedasticity as the weight variable.

- **Weight Variable.** The data are weighted by the reciprocal of this variable raised to a power. The regression equation is calculated for each of a specified range of power values and indicates the power that maximizes the log-likelihood function.

- **Power Range.** This is used in conjunction with the weight variable to compute weights. Several regression equations will be fit, one for each value in the power range. The values entered in the Power range test box and the through text box must be between -6.5 and 7.5, inclusive. The power values range from the low to high value, in increments determined by the value specified. The total number of values in the power range is limited to 150.

Weight Estimation Options

Figure 6-2
Weight Estimation Options dialog box

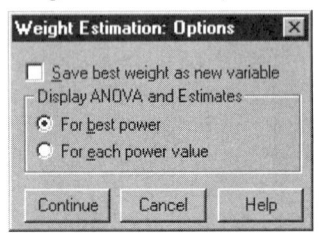

You can specify options for your weight estimation analysis:

Save best weight as new variable. Adds the weight variable to the active file. This variable is called *WGT_n*, where *n* is a number chosen to give the variable a unique name.

Display ANOVA and Estimates. Allows you to control how statistics are displayed in the output. Available alternatives are For best power and For each power value.

WLS Command Additional Features

The SPSS command language also allows you to:

- Provide a single value for the power.
- Specify a list of power values, or mix a range of values with a list of values for the power.

Chapter 7

Two-Stage Least-Squares Regression

Standard linear regression models assume that errors in the dependent variable are uncorrelated with the independent variable(s). When this is not the case (for example, when relationships between variables are bidirectional), linear regression using ordinary least squares (OLS) no longer provides optimal model estimates. Two-stage least-squares regression uses instrumental variables that are uncorrelated with the error terms to compute estimated values of the problematic predictor(s) (the first stage), and then uses those computed values to estimate a linear regression model of the dependent variable (the second stage). Since the computed values are based on variables that are uncorrelated with the errors, the results of the two-stage model are optimal.

Example. Is the demand for a commodity related to its price and consumers' incomes? The difficulty in this model is that price and demand have a reciprocal effect on each other. That is, price can influence demand and demand can also influence price. A two-stage least-squares regression model might use consumers' incomes and lagged price to calculate a proxy for price that is uncorrelated with the measurement errors in demand. This proxy is substituted for price itself in the originally specified model, which is then estimated.

Statistics. For each model: standardized and unstandardized regression coefficients, multiple R, R^2, adjusted R^2, standard error of the estimate, analysis-of-variance table, predicted values, and residuals. Also, 95% confidence intervals for each regression coefficient, and correlation and covariance matrices of parameter estimates.

Chapter 7

Two-Stage Least-Squares Regression Data Considerations

Data. The dependent and independent variables should be quantitative. Categorical variables, such as religion, major, or region of residence, need to be recoded to binary (dummy) variables or other types of contrast variables. **Endogenous** explanatory variables should be quantitative (not categorical).

Assumptions. For each value of the independent variable, the distribution of the dependent variable must be normal. The variance of the distribution of the dependent variable should be constant for all values of the independent variable. The relationship between the dependent variable and each independent variable should be linear.

Related procedures. If you believe that none of your predictor variables is correlated with the errors in your dependent variable, you can use the Linear Regression procedure. If your data appear to violate one of the assumptions (such as normality or constant variance), try transforming them. If your data are not related linearly and a transformation does not help, use an alternate model in the Curve Estimation procedure. If your dependent variable is dichotomous, such as whether a particular sale is completed or not, use the Logistic Regression procedure. If your data are not independent—for example, if you observe the same person under several conditions—use the Repeated Measures procedure, available in the SPSS Advanced Models option.

Obtaining a Two-Stage Least-Squares Regression Analysis

▶ From the menus choose:

Analyze
  Regression
    2-Stage Least Squares...

Two-Stage Least-Squares Regression

Figure 7-1
2-Stage Least Squares dialog box

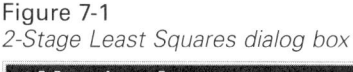

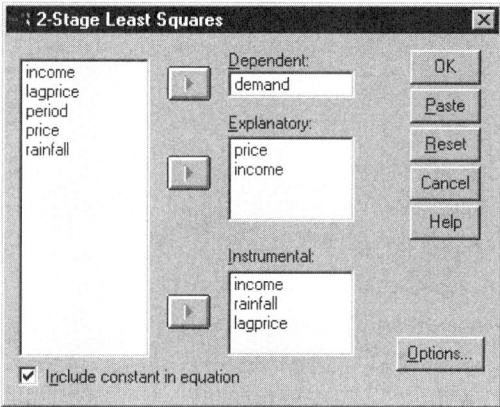

▶ Select one dependent variable.

▶ Select one or more explanatory (predictor) variables.

▶ Select one or more instrumental variables.

- **Instrumental.** These are the variables used to compute the predicted values for the endogenous variables in the first stage of two-stage least squares analysis. The same variables may appear in both the Explanatory and Instrumental list boxes. The number of instrumental variables must be at least as many as the number of explanatory variables. If all explanatory and instrumental variables listed are the same, the results are the same as results from the Linear Regression procedure.

Explanatory variables not specified as instrumental are considered endogenous. Normally, all of the exogenous variables in the Explanatory list are also specified as instrumental variables.

Two-Stage Least-Squares Regression Options

Figure 7-2
2-Stage Least Squares Options dialog box

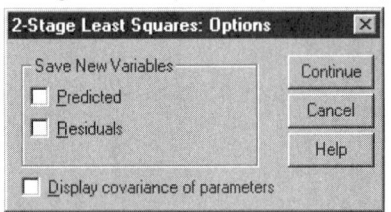

You can select the following options for your analysis:

Save New Variables. Allows you to add new variables to your active file. Available options are Predicted and Residuals.

Display covariance of parameters. Allows you to print the covariance matrix of the parameter estimates.

2SLS Command Additional Features

The SPSS command language also allows you to estimate multiple equations simultaneously.

Appendix

A

Categorical Variable Coding Schemes

In many SPSS procedures, you can request automatic replacement of a categorical independent variable with a set of contrast variables, which will then be entered or removed from an equation as a block. You can specify how the set of contrast variables is to be coded, usually on the CONTRAST subcommand. This appendix explains and illustrates how different contrast types requested on CONTRAST actually work.

Deviation

Deviation from the grand mean. In matrix terms, these contrasts have the form:

```
mean      ( 1/k       1/k      ...    1/k      1/k  )
df(1)     ( 1–1/k    –1/k      ...   –1/k     –1/k  )
df(2)     ( –1/k    1–1/k      ...   –1/k     –1/k  )
              .         .
              .         .
df(k–1)   ( –1/k     –1/k      ...  1–1/k     –1/k  )
```

where k is the number of categories for the independent variable and the last category is omitted by default. For example, the deviation contrasts for an independent variable with three categories are as follows:

```
( 1/3       1/3      1/3 )
( 2/3      –1/3     –1/3 )
(–1/3       2/3     –1/3 )
```

51

Appendix A

To omit a category other than the last, specify the number of the omitted category in parentheses after the DEVIATION keyword. For example, the following subcommand obtains the deviations for the first and third categories and omits the second:

/CONTRAST(FACTOR)=DEVIATION(2)

Suppose that *factor* has three categories. The resulting contrast matrix will be

$$\begin{pmatrix} 1/3 & 1/3 & 1/3 \\ 2/3 & -1/3 & -1/3 \\ -1/3 & -1/3 & 2/3 \end{pmatrix}$$

Simple

Simple contrasts. Compares each level of a factor to the last. The general matrix form is

mean	(1/k	1/k	...	1/k	1/k)
df(1)	(1	0	...	0	−1)
df(2)	(0	1	...	0	−1)
.		.			
.		.			
df(k−1)	(0	0	...	1	−1)

where k is the number of categories for the independent variable. For example, the simple contrasts for an independent variable with four categories are as follows:

$$\begin{pmatrix} 1/4 & 1/4 & 1/4 & 1/4 \\ 1 & 0 & 0 & -1 \\ 0 & 1 & 0 & -1 \\ 0 & 0 & 1 & -1 \end{pmatrix}$$

To use another category instead of the last as a reference category, specify in parentheses after the SIMPLE keyword the sequence number of the reference category, which is not necessarily the value associated with that category. For example, the following CONTRAST subcommand obtains a contrast matrix that omits the second category:

/CONTRAST(FACTOR) = SIMPLE(2)

Suppose that *factor* has four categories. The resulting contrast matrix will be

$$\begin{pmatrix} 1/4 & 1/4 & 1/4 & 1/4 \\ 1 & -1 & 0 & 0 \\ 0 & -1 & 1 & 0 \\ 0 & -1 & 0 & 1 \end{pmatrix}$$

Helmert

Helmert contrasts. Compares categories of an independent variable with the mean of the subsequent categories. The general matrix form is

$$\begin{array}{l} \text{mean} \\ df(1) \\ df(2) \\ \\ \\ df(k-2) \\ df(k-1) \end{array} \begin{pmatrix} 1/k & 1/k & \cdots & 1/k & 1/k \\ 1 & -1/(k-1) & \cdots & -1/(k-1) & -1/(k-1) \\ 0 & 1 & \cdots & -1/(k-2) & -1/(k-2) \\ \vdots & & \vdots & & \\ 0 & 0 & 1 & -1/2 & -1/2 \\ 0 & 0 & \cdots & 1 & -1 \end{pmatrix}$$

where k is the number of categories of the independent variable. For example, an independent variable with four categories has a Helmert contrast matrix of the following form:

$$\begin{pmatrix} 1/4 & 1/4 & 1/4 & 1/4 \\ 1 & -1/3 & -1/3 & -1/3 \\ 0 & 1 & -1/2 & -1/2 \\ 0 & 0 & 1 & -1 \end{pmatrix}$$

Appendix A

Difference

Difference or reverse Helmert contrasts. Compares categories of an independent variable with the mean of the previous categories of the variable. The general matrix form is

$$\begin{array}{lcccccc}
\text{mean} & (\ 1/k & 1/k & 1/k & \cdots & 1/k\ ) \\
df(1) & (-1 & 1 & 0 & \cdots & 0\ ) \\
df(2) & (-1/2 & -1/2 & 1 & \cdots & 0\ ) \\
& \cdot & \cdot & & & \\
& \cdot & \cdot & & & \\
df(k-1) & (-1/(k-1) & -1/(k-1) & -1/(k-1) & \cdots & 1\ )
\end{array}$$

where k is the number of categories for the independent variable. For example, the difference contrasts for an independent variable with four categories are as follows:

$$\begin{array}{cccc}
(\ 1/4 & 1/4 & 1/4 & 1/4\ ) \\
(-1 & 1 & 0 & 0\ ) \\
(-1/2 & -1/2 & 1 & 0\ ) \\
(-1/3 & -1/3 & -1/3 & 1\ )
\end{array}$$

Polynomial

Orthogonal polynomial contrasts. The first degree of freedom contains the linear effect across all categories; the second degree of freedom, the quadratic effect; the third degree of freedom, the cubic; and so on, for the higher-order effects.

You can specify the spacing between levels of the treatment measured by the given categorical variable. Equal spacing, which is the default if you omit the metric, can be specified as consecutive integers from 1 to k, where k is the number of categories. If the variable *drug* has three categories, the subcommand

/CONTRAST(DRUG)=POLYNOMIAL

is the same as

/CONTRAST(DRUG)=POLYNOMIAL(1,2,3)

Categorical Variable Coding Schemes

Equal spacing is not always necessary, however. For example, suppose that *drug* represents different dosages of a drug given to three groups. If the dosage administered to the second group is twice that given to the first group and the dosage administered to the third group is three times that given to the first group, the treatment categories are equally spaced, and an appropriate metric for this situation consists of consecutive integers:

/CONTRAST(DRUG)=POLYNOMIAL(1,2,3)

If, however, the dosage administered to the second group is four times that given to the first group, and the dosage administered to the third group is seven times that given to the first group, an appropriate metric is

/CONTRAST(DRUG)=POLYNOMIAL(1,4,7)

In either case, the result of the contrast specification is that the first degree of freedom for *drug* contains the linear effect of the dosage levels and the second degree of freedom contains the quadratic effect.

Polynomial contrasts are especially useful in tests of trends and for investigating the nature of response surfaces. You can also use polynomial contrasts to perform nonlinear curve fitting, such as curvilinear regression.

Repeated

Compares adjacent levels of an independent variable. The general matrix form is

mean	($1/k$	$1/k$	$1/k$	...	$1/k$	$1/k$)
df(1)	(1	−1	0	...	0	0)
df(2)	(0	1	−1	...	0	0)
.		.				
.		.				
df(k−1)	(0	0	0	...	1	−1)

where k is the number of categories for the independent variable. For example, the repeated contrasts for an independent variable with four categories are as follows:

Appendix A

```
( 1/4   1/4   1/4   1/4 )
(  1    -1    0     0  )
(  0     1   -1     0  )
(  0     0    1    -1  )
```

These contrasts are useful in profile analysis and wherever difference scores are needed.

Special

A user-defined contrast. Allows entry of special contrasts in the form of square matrices with as many rows and columns as there are categories of the given independent variable. For MANOVA and LOGLINEAR, the first row entered is always the mean, or constant, effect and represents the set of weights indicating how to average other independent variables, if any, over the given variable. Generally, this contrast is a vector of ones.

The remaining rows of the matrix contain the special contrasts indicating the desired comparisons between categories of the variable. Usually, orthogonal contrasts are the most useful. Orthogonal contrasts are statistically independent and are nonredundant. Contrasts are orthogonal if:

- For each row, contrast coefficients sum to 0.
- The products of corresponding coefficients for all pairs of disjoint rows also sum to 0.

For example, suppose that treatment has four levels and that you want to compare the various levels of treatment with each other. An appropriate special contrast is

```
( 1   1   1   1 )    weights for mean calculation
( 3  -1  -1  -1 )    compare 1st with 2nd through 4th
( 0   2  -1  -1 )    compare 2nd with 3rd and 4th
( 0   0   1  -1 )    compare 3rd with 4th
```

which you specify by means of the following CONTRAST subcommand for MANOVA, LOGISTICREGRESSION, and COXREG:

Categorical Variable Coding Schemes

```
/CONTRAST(TREATMNT)=SPECIAL( 1  1  1  1
                             3 -1 -1 -1
                             0  2 -1 -1
                             0  0  1 -1 )
```

For LOGLINEAR, you need to specify:

```
/CONTRAST(TREATMNT)=BASIS SPECIAL( 1  1  1  1
                                   3 -1 -1 -1
                                   0  2 -1 -1
                                   0  0  1 -1 )
```

Each row except the means row sums to 0. Products of each pair of disjoint rows sum to 0 as well:

Rows 2 and 3: $(3)(0) + (-1)(2) + (-1)(-1) + (-1)(-1) = 0$

Rows 2 and 4: $(3)(0) + (-1)(0) + (-1)(1) + (-1)(-1) = 0$

Rows 3 and 4: $(0)(0) + (2)(0) + (-1)(1) + (-1)(-1) = 0$

The special contrasts need not be orthogonal. However, they must not be linear combinations of each other. If they are, the procedure reports the linear dependency and ceases processing. Helmert, difference, and polynomial contrasts are all orthogonal contrasts.

Indicator

Indicator variable coding. Also known as dummy coding, this is not available in LOGLINEAR or MANOVA. The number of new variables coded is $k-1$. Cases in the reference category are coded 0 for all $k-1$ variables. A case in the ith category is coded 0 for all indicator variables except the ith, which is coded 1.

Index

asymptotic regression, 35
  in Nonlinear Regression, 35
backward elimination, 6
  in Logistic Regression, 6
binary logistic regression, 1
categorical covariates, 7
cell probabilities tables, 18
  in Multinomial Logistic Regression, 18
cells with zero observations, 19
  in Multinomial Logistic Regression, 19
classification, 13
  in Multinomial Logistic Regression, 13
classification tables, 18
  in Multinomial Logistic Regression, 18
confidence intervals, 18
  in Multinomial Logistic Regression, 18
constant term, 10
  in Linear Regression, 10
constrained regression
  in Nonlinear Regression, 37
contrasts, 7
  in Logistic Regression, 7
convergence criterion, 19
  in Multinomial Logistic Regression, 19
Cook's D, 9
  in Logistic Regression, 9
correlation matrix, 18
  in Multinomial Logistic Regression, 18
covariance matrix, 18
  in Multinomial Logistic Regression, 18
covariates
  in Logistic Regression, 7
Cox and Snell R-square, 18
  in Multinomial Logistic Regression, 18
custom models, 15
  in Multinomial Logistic Regression, 15

delta, 19
  as correction for cells with zero observations, 19
density model, 35
  in Nonlinear Regression, 35
deviance function, 21
  for estimating dispersion scaling value, 21

DfBeta, 9
  in Logistic Regression, 9
dispersion scaling value, 21
  in Multinomial Logistic Regression, 21
fiducial confidence intervals, 28
  in Probit Analysis, 28
forward selection, 6
  in Logistic Regression, 6
full factorial models, 15
  in Multinomial Logistic Regression, 15
Gauss model, 35
  in Nonlinear Regression, 35
Gompertz model, 35
  in Nonlinear Regression, 35
goodness of fit, 18
  in Multinomial Logistic Regression, 18
Hosmer-Lemeshow goodness-of-fit statistic, 10
  in Logistic Regression, 10
intercept, 15
  include or exclude, 15
iteration history, 19
  in Multinomial Logistic Regression, 19
iterations, 10, 19, 28
  in Logistic Regression, 10
  in Multinomial Logistic Regression, 19
  in Probit Analysis, 28
Johnson-Schumacher model, 35
  in Nonlinear Regression, 35
leverage values, 9
  in Logistic Regression, 9
likelihood ratio, 18, 21
  for estimating dispersion scaling value, 21
  goodness of fit, 18
Linear Regression, 43, 47
  Two-Stage Least-Squares Regression, 47
  weight estimation, 43
logistic regression, 1
  binary, 1
Logistic Regression, 3, 6
  assumptions, 4
  categorical covariates, 7

Index

classification cutoff, 10
coefficients, 3
command additional features, 11
constant term, 10
contrasts, 7
data considerations, 4
define selection rule, 6
dependent variable, 4
display options, 10
example, 3
Hosmer-Lemeshow goodness-of-fit statistic, 10
influence measures, 9
iterations, 10
predicted values, 9
probability for stepwise, 10
related procedures, 4
residuals, 9
saving new variables, 9
set rule, 6
statistics, 3
statistics and plots, 10
string covariates, 7
variable selection methods, 6
log-likelihood, 18, 43
in Multinomial Logistic Regression, 18
in Weight Estimation, 43
log-modified model, 35
in Nonlinear Regression, 35

main-effects models, 15
in Multinomial Logistic Regression, 15
McFadden R-square, 18
in Multinomial Logistic Regression, 18
Metcherlich law of diminishing returns, 35
in Nonlinear Regression, 35
Michaelis Menten model, 35
in Nonlinear Regression, 35
Morgan-Mercer-Florin model, 35
in Nonlinear Regression, 35
Multinomial Logistic Regression, 13, 13, 15, 18, 19, 22, 23
assumptions, 13
command additional features, 23
criteria, 19
exporting model information, 22
models, 15

reference category, 17
save, 22
statistics, 18
Nagelkerke R-square, 18
in Multinomial Logistic Regression, 18
nonlinear models, 35
in Nonlinear Regression, 35
Nonlinear Regression, 31, 35
assumptions, 32
bootstrap estimates, 39
command additional features, 40
common nonlinear models, 35
conditional logic, 33
data considerations, 32
derivatives, 38
estimation methods, 39
example, 31
interpretation of results, 40
Levenberg-Marquardt algorithm, 39
loss function, 36
parameter constraints, 37
parameters, 34
predicted values, 38
related procedures, 32
residuals, 38
save new variables, 38
segmented model, 33
sequential quadratic programming, 39
starting values, 32, 34
statistics, 31

parallelism test, 28
in Probit Analysis, 28
parameter constraints
in Nonlinear Regression, 37
parameter estimates, 18
in Multinomial Logistic Regression, 18
Peal-Reed model, 35
in Nonlinear Regression, 35
Pearson chi-square, 18, 21
for estimating dispersion scaling value, 21
goodness of fit, 18
Probit Analysis, 25
assumptions, 25
command additional features, 29
criteria, 28

Index

data considerations, 25
define range, 28
example, 25
fiducial confidence intervals, 28
iterations, 28
natural response rate, 28
parallelism test, 28
related procedures, 25
relative median potency, 28
statistics, 25, 28

ratio of cubics model, 35
  in Nonlinear Regression, 35
ratio of quadratics model, 35
  in Nonlinear Regression, 35
reference category
  in Multinomial Logistic Regression, 17
relative median potency, 28
  in Probit Analysis, 28
Richards model, 35
  in Nonlinear Regression, 35

separation, 19
  in Multinomial Logistic Regression, 19
singularity, 19
  in Multinomial Logistic Regression, 19
step-halving, 19
  in Multinomial Logistic Regression, 19
stepwise selection, 6, 15
  in Logistic Regression, 6
  in Multinomial Logistic Regression, 15
string covariates, 7
  in Logistic Regression, 7

Two-Stage Least-Squares Regression, 47, 48, 50, 50
  assumptions, 48
  command additional features, 50
  covariance of parameters, 50
  data considerations, 48
  example, 47
  instrumental variables, 47
  related procedures, 48
  saving new variables, 50
  statistics, 47

Verhulst model, 35
  in Nonlinear Regression, 35
Von Bertalanffy model, 35
  in Nonlinear Regression, 35
Weibull model, 35
  in Nonlinear Regression, 35
Weight Estimation, 43
  assumptions, 43
  command additional features, 46
  data considerations, 43
  display ANOVA and estimates, 46
  example, 43
  iteration history, 46
  log-likelihood, 43
  related procedures, 43
  save best weights as new variable, 46
  statistics, 43

yield density model, 35
  in Nonlinear Regression, 35